Detectives: Animal Smarts

Search for the Facts...

Builders

Anne O'Daly

Published by Brown Bear Books Ltd
4877 N. Circulo Bujia
Tucson, AZ 85718
USA

and

G14, Regent Studios
1 Thane Villas
London N7 7PH
UK

ISBN 978-1-83572-023-3 (ALB)
ISBN 978-1-83572-029-5 (paperback)
ISBN 978-1-83572-035-6 (ebook)

Library of Congress Cataloging-in-Publication Data available on request

Design Manager: Keith Davis
Children's Publisher: Anne O'Daly
Picture Manager: Sophie Mortimer

Picture Credits
Cover: iStock: MyImages_Micha. Interior: iStock: Anita Andre 16–17, Bruce Block 10–11, Gergory Dubus 21, Vini Souza 12, Boshoff Steenekamp 5c; Shutterstock: Georgi Baird 14–15, Volodymyr Burdiak 22bl, Adrian Eugen Cioabaniuc 10, Heather M Davidson 16, Ian Fletcher 4–5t, Ghost Bear 5tl, Simon Greig 12–13, Bob Hilscher 6–7, Cathy Keifer 22tr, Isabel Kendzior 5b, 23, Tools Konten 5tr, Gleb Korovko 4t, Vera Larina 22br, Elina Litovkina 20, Dan Nurgitz 22tl, Caryn Pereira 8–9, Vadim Serebrenikov 1, 4b, Debbie Steinhausser 18, Kuttelvaserova Stuchelova 18–19, Alessandro Tortora 14, Vladimir Turkenich 7, Maximiliane Wagner 9, Apisit Wilaijit 4–5b.
t=top, b=bottom, l=left, r=right, c=center
All artwork and other photography Brown Bear Books.

Brown Bear Books has made every attempt to contact the copyright holder.
If you have any information about omissions, please contact: licensing@brownbearbooks.co.uk

Manufactured in the United States of America
CPSIA compliance information: Batch#AG/5663

Websites
The website addresses in this book were valid at the time of going to press. However, it is possible that contents or addresses may change following publication of this book. No responsibility for any such changes can be accepted by the author or the publisher. Readers should be supervised when they access the Internet.

Contents

Meet the Builders

Animals are clever builders. Some build homes for shelter. Others make traps to catch prey. Meet some amazing builders!

Some weaverbirds spend 12 days making their nest.

Beaver

Beavers build stick homes called lodges. They make them in pools. Part of the lodge is under water. If the pool isn't deep enough, the beavers make it deeper!

FACT FILE

Name: beaver

Where it lives: near rivers, streams, and lakes

What it eats: leaves, twigs, and tree bark

Type of animal: mammal

Size: 3 to 4 feet (0.9 to 1.2 m)

Body: strong teeth and jaws; thick tail for swimming

Beavers make their home with sticks and mud

INSIDE A BEAVER LODGE

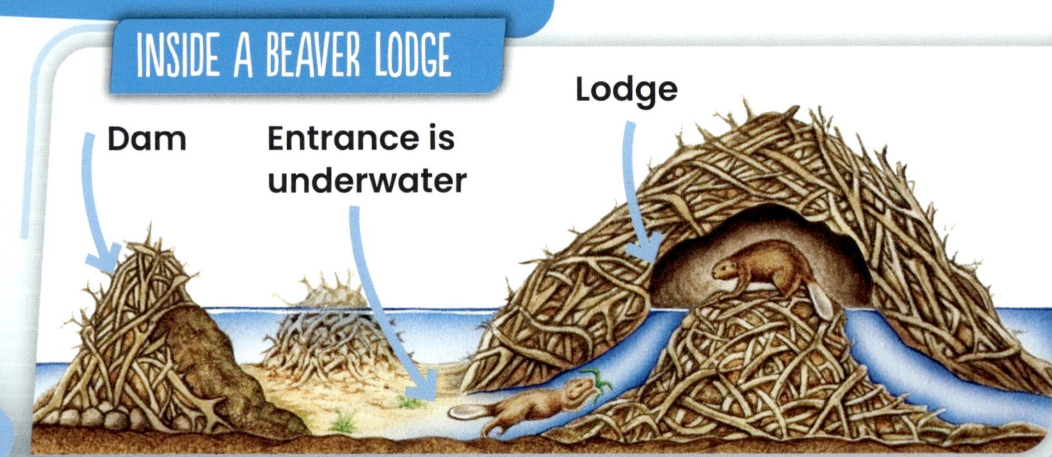

They put logs, grass, and leaves on top

The dam makes a deep pond

Beavers have strong teeth for biting wood. They can cut down a tree in less than five minutes.

Weaverbird

Weaverbirds make amazing nests.
The males build a nest to attract a mate.
They collect grass and leaves.
The birds weave them into a ball.

FACT FILE

Name: Eastern golden weaver

Where it lives: near rivers and coasts in eastern Africa

What it eats: seeds, nectar, and insects

Type of animal: bird

Size: 5.9 inches (15 cm)

Body: yellow feathers; male has a black bill and red eyes

The bird ties the nest to the tree

The entrance is at the bottom

MINI FACTS

One nest can have up to 1,000 pieces of grass in it.

He holds the grass with his foot

Females pick the best nest

Some weavers build massive nests. The biggest can fit 500 birds. Each family has its own room.

Spider

Many spiders make sticky webs. They spin them from strong silk. Insects fly into the web. They get stuck. The spider has caught a tasty meal!

FACT FILE

Name: garden orb weavers

Where it lives: fields, gardens, and forests around the world

What it eats: insects

Type of animal: arachnid

Size: females 0.7 to 1.2 inches (1.8 to 3 cm), males 0.6 to 0.7 inches (1.5 to 1.8 cm)

Body: brown, yellow, or black body; some have a stripe along their back

The web is made from silk threads

The spider makes the silk inside its body

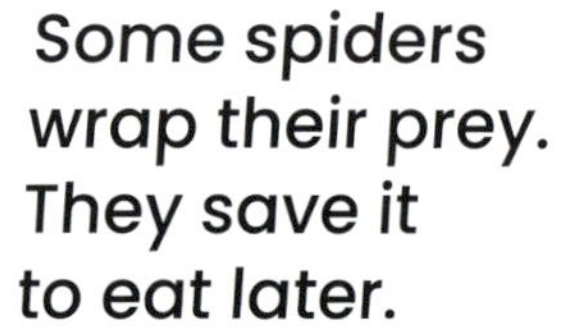

Some spiders wrap their prey. They save it to eat later.

The spider waits in the middle of the web

MINI FACTS

Spider silk is strong and stretchy. It's five times stronger than steel!

Termite

Termites are tiny insects.
But they build big homes.
The homes are called mounds.
The mounds have tunnels inside.

FACT FILE

Name: mound-building termite

Where it lives: Africa, South America, and Australia

What it eats: grasses and dry leaves

Type of animal: insect

Size: 0.16 to 0.6 inches (4 to 15 mm)

Body: workers have a brown body; soldiers have strong jaws to fight off enemies

Termites use soil to make mounds. They mix it with saliva and poop!

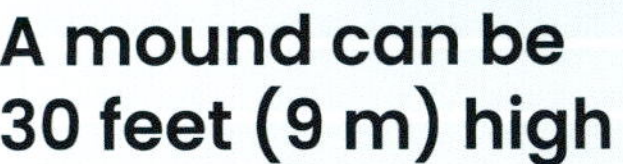

A mound can be 30 feet (9 m) high

Millions of termites live in the mound

Tunnels let air get into the mound. They keep it cool.

MINI FACTS

Termites live in a colony. They all have jobs. Workers collect food. Soldiers guard the colony. A king and queen rule the colony.

Prairie Dog

These furry rodents are built for digging! They dig burrows under the prairies in North America. Some live in underground towns.

FACT FILE

Name: prairie dog

Where it lives: grasslands of North America

What it eats: mainly grasses and dry leaves

Type of animal: mammal

Size: up to 13 inches (33 cm)

Body: brown fur, short tail, round head with small ears

Prairie dogs rub noses when they meet.

A prairie dog looks out for danger

Strong feet and sharp claws for digging

Babies come out of the burrow after 6 weeks

INSIDE A PRAIRIE DOG TOWN

The burrow has rooms for different activities.

Toilet

Foodstore

Bedroom

Polar Bear

Polar bears live in the Arctic.
It is freezing all year round.
A female bear digs a den.
Her cubs are born there.

FACT FILE

Name: polar bear

Where it lives: the Arctic, on sea ice and on the shores

What it eats: mainly seals

Type of animal: mammal

Size: up to 8 feet (2.4 m)

Body: thick fur to keep warm; strong, wide paws for swimming

The mother digs the den in the fall

Polar bears have furry paws. They have strong claws for digging.

MINI FACTS

There is no food in the den. The mother lives off fat in her body. The cubs feed on their mother's milk.

In the spring, the bears leave the den

The den keeps the family safe and warm

Honeybee

Honeybees live in nests called hives. They make a sticky wax to build the hive. The bees make rooms called cells.

FACT FILE

Name: honeybee

Where it lives: gardens, woodlands, and meadows around the world

What it eats: nectar and pollen

Type of animal: insect

Size: around 0.5 in (1.3 cm) long

Body: yellow body with brown stripes, six legs

More than 20,000 bees live in the hive

Many bees make nests in caves or trees.

Some cells have eggs inside

Each cell has six sides

Workers store food in the cells

MINI FACTS

The hive has three types of bees. The queen rules the hive. She lays eggs. Workers make the hive. They fix it and keep it clean. Drones mate with the queen.

Ant Lion

Ant lions are clever hunters.
They dig a pit in sand.
Then they wait at the bottom.
If an insect falls in, the ant lion eats it.

FACT FILE

Name: ant lion

Where it lives: dry, sandy places around the world

What it eats: insects

Type of animal: insect

Size: larvae are 0.5 in (1.3 cm)

Body: the larvae have plump bodies and large jaws

The pit is 1.2 inches (3 cm) deep

The prey can't climb up the slippery side

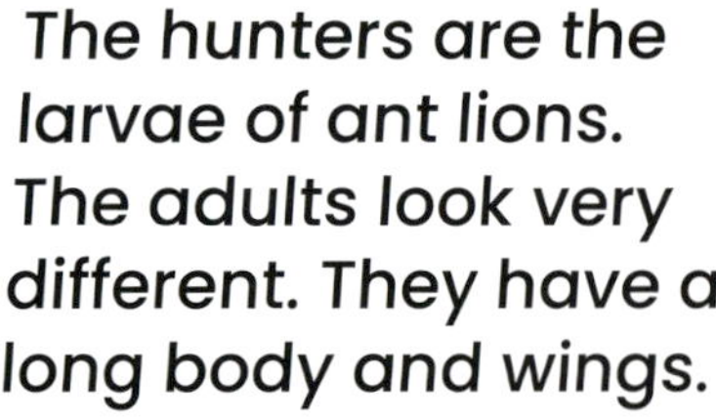

The hunters are the larvae of ant lions. The adults look very different. They have a long body and wings.

MINI FACTS

Ant lions are also called doodlebugs. They make squiggly marks in the sand. The marks look like doodles.

The ant lion grabs the prey with its strong jaws

Quiz

Test your skills!
Can you answer these questions?
The answers are on page 24.

1. What is a beaver's home called?

2. Where do spiders make silk?

3. What is a termite mound made from?

4. What does a queen honeybee do?

Useful Words

arachnid A type of animal with eight legs. Spiders are arachnids.

Arctic The area around the North Pole. It is icy and cold all year round.

colony A group of animals of the same kind that live together.

honeycomb A collection of cells made by bees. Each cell has thin walls and six sides.

insect An animal with hard body and six legs. Honeybees and termites are insects.

larvae The young of certain animals. They look very different from the adults.

mammal An animal that feeds its young on milk. Most mammals have fur or hair.

prey An animal that is hunted by other animals for food.

Find Out More

Books

Animal Builders (Spot Best Ever Animals), Golriz Golkar (Amicus Learning, 2024)

Animal Engineers (Astonishing Animals), Izzi Howell (Crabtree Publishing, 2020)

Beavers (Animal Engineers), Rebecca Pettiford (Blastoff! Readers, 2024)

Websites

www.bbc.co.uk/programmes/articles/52z8GYfsRsz1JZt2gz10zvd/natures-top-five-best-builders

www.dawn.com/news/1469837

www.globetrottinkids.com/wp-content/uploads/2022/01/animal-architects.pdf

Index

Quiz Answers: 1. It's called a lodge; **2.** They make silk inside their bodies; **3.** It's made from soil, saliva, and poop; **4.** The queen honeybee lays eggs.